WIND

A Crabtree Roots Book

DOUGLAS BENDER

School-to-Home Support for Caregivers and Teachers

This book helps children grow by letting them practice reading. Here are a few guiding questions to help the reader with building his or her comprehension skills. Possible answers appear here in red.

Before Reading:

- What do I think this book is about?
  - *I think this book is about how wind is made.*
  - *I think this book is about how wind moves.*

- What do I want to learn about this topic?
  - *I want to learn what wind is made up of.*
  - *I want to learn how the wind helps make things fly.*

During Reading:

- I wonder why...
  - *I wonder why we can't see wind.*
  - *I wonder why we can hear wind.*

- What have I learned so far?
  - *I have learned that hot and cold air create wind.*
  - *I have learned that wind helps people fly kites.*

After Reading:

- What details did I learn about this topic?
  - *I have learned that waves are created by wind.*
  - *I have learned that wind is made up of air.*

- Read the book again and look for the vocabulary words.
  - *I see the word* ***hot*** *on page 6 and the word* ***kite*** *on page 11. The other vocabulary words are found on page 14.*

Whoosh! Can you hear the wind?

Wind is made up of air.

Hot air moves up to the sky.

Cold air moves down to where land is.

This is what makes wind move all around us.

Wind helps us fly a **kite**!

Wind can also turn water into **waves**.

Word List

Sight Words

a
all
around
can
down
fly
helps
into
is
made
makes
of
the
this
to
up
us
water
wind
what
where
you

Words to Know

cold

hot

kite

waves

49 Words

Whoosh! Can you hear the wind?

Wind is made up of air.

Hot air moves up to the sky.

Cold air moves down to where land is.

This is what makes wind move all around us.

Wind helps us fly a **kite**!

Wind can also turn water into **waves**.

Written by: Douglas Bender
Designed by: Rhea Wallace
Series Development: James Earley
Proofreader: Janine Deschenes
Educational Consultant: Marie Lemke M.Ed.

Photographs:
Shutterstock: Theodore Trimmer: cover; Kateryna Mashkevych: p. 1; SingerGM: p. 3; Brian A. Jackson: p. 5; Vadym Laura: p. 6-7, 14; Chalermpon Poungpeth: p. 9; Olesia Bilkei: p. 10, 14; Rileysmithphotography: p. 12, 14

Library and Archives Canada Cataloguing in Publication
Title: Wind / Douglas Bender.
Names: Bender, Douglas, 1992- author.
Description: Series statement: The weather forecast | "A Crabtree roots book".
Identifiers: Canadiana (print) 20210181370 | Canadiana (ebook) 20210181389 | ISBN 9781427159373 (hardcover) | ISBN 9781427159434 (softcover) | ISBN 9781427133755 (HTML) | ISBN 9781427134356 (EPUB) | ISBN 9781427159618 (read-along ebook)
Subjects: LCSH: Winds—Juvenile literature.
Classification: LCC QC931.4 .B46 2022 | DDC j551.51/8—dc23

Library of Congress Cataloging-in-Publication Data
Names: Bender, Douglas, 1992- author.
Title: Wind / Douglas Bender.
Description: New York : Crabtree Publishing, [2022] | Series: The weather forecast - a Crabtree roots book | Includes index.
Identifiers: LCCN 2021014525 (print) | LCCN 2021014526 (ebook) | ISBN 9781427159373 (hardcover) | ISBN 9781427159434 (paperback) | ISBN 9781427133755 (ebook) | ISBN 9781427134356 (epub) | ISBN 9781427159618
Subjects: LCSH: Winds--Juvenile literature.
Classification: LCC QC931.4 .B45 2022 (print) | LCC QC931.4 (ebook) | DDC 551.51/8--dc23
LC record available at https://lccn.loc.gov/2021014525
LC ebook record available at https://lccn.loc.gov/2021014526

Crabtree Publishing Company
www.crabtreebooks.com 1-800-387-7650

Printed in Canada/042022/CPC20220411

 In Canada: We acknowledge the financial support of the Government of Canada through the Canada Book Fund for our publishing activities.

Published in the United States
Crabtree Publishing
347 Fifth Avenue, Suite 1402-145
New York, NY, 10016

Published in Canada
Crabtree Publishing
616 Welland Ave.
St. Catharines, Ontario L2M 5V6